ANIMAL ENCYCLOPEDIA

Whales & Dolphins

MACAW
BOOKS

ANIMAL ENCYCLOPEDIA

Whales & Dolphins

First published in 2008

Published by
Macaw Books
www.macawbooks.com

ISBN-978-1-60346-133-7

Printed in India

Contents

Whales and Dolphins

Whales and dolphins are large, carnivorous, marine mammals. They are related to each other under the group "cetaceans." There are about 80 species of whales and dolphins in the world.

Mammals

Whales and dolphins resemble fish in many ways, but they are actually mammals, being warm-blooded animals that give birth to live young and nurse them, just like other mammals. Unlike fish, which use gills to breathe and have scales covering their body, whales and dolphins have hair on their body and breathe through blowholes using their lungs.

Evolution

Whales and dolphins are believed to have evolved in the Cretaceous Period from large, terrestrial mammals that belonged to the family of Mesonychidae. One of the earliest known whales was ambulocetus, which looked like a furry crocodile and could swim as well as move on land.

Baleen Whales

Whales are classified into two groups: baleen whales and toothed whales. Baleen whales do not have teeth, but rather, have hundreds of thin plates with brush-like fibers, which hang from their upper jaw and are known as baleen or whalebone. Baleen whales use these plates to filter plankton and other microscopic organisms from water for food. There are three types of baleen whales: right whales, gray whales, and rorquals.

Toothed Whales

Toothed whales include sperm whales, belugas, beaked whales, dolphins, and river dolphins. These whales have hundreds of teeth but do not use them to feed on prey, but rather to catch hold of it. Once they have their prey, toothed whales swallow it whole.

Blubber

Whales have a thick layer of fat beneath their skin, called blubber. Blubber protects them from the cold waters of their habitats, and at times, when food is scarce, whales can live off of their blubber for long periods.

Group Living

Whales and dolphins live in groups known as herds, pods, or schools. Many toothed whales, like pilot whales and some species of dolphins, live in groups of hundreds to thousands of individuals. Baleen whales generally live alone or in small groups but sometimes gather in large groups for feeding and breeding.

Beluga Whale

The beluga whale is a small, white, toothed whale around 3 to 5 meters long and weighing about 1,500 kilograms, smaller than most other toothed whales.

Beluga whales are also known as white whales. They are often hunted for their meat for human consumption and for their oil for use in cooking and as fuel. The oil is extracted from the blubber, which is a thick layer of fat beneath the skin that helps them to keep warm in the icy waters of the oceans.

Habitat and Range

Beluga whales are Arctic dwellers, living in the Arctic and sub-arctic waters of Russia, Canada, Norway, Alaska, and Greenland. These whales generally inhabit channels, bays, inlets, and shallow waters but sometimes are also found at river mouths in summertime for feeding, socializing, and birthing.

Calves

Beluga whales give birth to a single calf every two to three years and do so in warm, shallow, freshwater rivers. Calves are dark gray and becomes lighter as they mature. A newborn calf is 1.2 to 1.5 meters long and weighs between 50 and 60 kilograms.

Did You Know?

There are approximately 62,000 to 80,000 beluga whales in the world.

Beluga herds

Beluga whales are highly social and playful animals who are often seen chasing or rubbing against each other. Beluga whales move and hunt in large groups generally made up of individuals of the same age and gender. Mother belugas move in smaller groups with their calves.

Communication

Beluga whales are the most vocal of all whales and dolphins and are able to make a wide range of sounds. Their loud, bird-like chirps and whistles have earned them the nickname "sea canaries." They also use body language, such as splashing around and grinding their jaws, to communicate with each other.

Anatomy

Beluga whales have a stocky, cigar-shaped body and a round, melon-shaped head with a flexible neck and a short snout. These whales have a thick layer of blubber which keeps them warm in the icy waters of the Arctic Ocean. Beluga whales do not have a dorsal fin but instead have a narrow ridge that runs along their back. Belugas have short, broad flippers and a broad fluke; the flippers are curled up at the tips, and the fluke has a deep indentation down the middle.

Preys and Predators

Beluga whales mostly prey on crab, snails, squid, and octopus; however, they also eat many species of fish including salmon, cod, smelt, flatfish, flounder, and sculpin. Beluga whales generally swallow their prey whole since they do not have very sharp teeth. Belugas are predated upon by killer whales and polar bears.

Blue Whale

The blue whale is the largest animal on Earth, measuring 21 to 24 meters in length and weighing 90,000 to 150,000 kilograms. Female blue whales are larger than the males.

Blue whales are blue-gray baleen whales with light gray markings. Their bellies often have a yellowish tinge because of the large number of microorganisms, called diatoms, that remain attached to their skin. Blue whales have about 300 to 400 baleen plates which they use to strain food from the ocean water.

Blue Whale Groups

Blue whales generally travel alone or in small groups of two to four individuals and are rarely seen in pairs. At times, large gatherings of blue whales occur near feeding grounds.

Anatomy

Blue whales have a tapered and streamlined body with a broad, flat, U-shaped head. Their skin is generally smooth, being encrusted with only a few barnacles and diatoms. Blue whales have twin blowholes that are each surrounded by large splashguards. They also have a ridge across the top of their head that extends to the tip of the snout.

Did You Know?

Blue whales are strong and fast swimmers and can travel at speeds of 30 miles per hour when alarmed.

Fins, Flippers, and Flukes

Blue whales have small dorsal fins which differ in shape from one individual to another. They have long, slender flippers with pointed tips and broad, triangular flukes.

Communication

Blue whales are the loudest animals in the world. They make low frequency, deep rumbling sounds to communicate with each other, and these sound waves travel across hundreds of miles through the ocean. Blue whales also slap their tails on the water's surface to convey messages to one another.

Habitat and Range

Blue whales are found in oceans throughout the world and prefer polar, temperate, and tropical waters during different seasons. In winter, they migrate to tropical-temperate waters and subtropical zones to breed and give birth to calves, and in the spring and summer, they spend their time in the polar regions.

Calves

Blue whales give birth to one calf every two to three years. Calves are around seven meters long at birth and weigh about 2,000 kilograms.

Food Habits

Blue whales are filter feeders. During feeding, they take in large volumes of water and food into their mouth, and as they close their mouth, the water is expelled through the baleen plates, and the food remains inside. Blue whales primarily feed on krill, eating up to four tons everyday—that is 40 million krill in a day!

Bottlenose Dolphin

The bottlenose dolphin is the most common dolphin in the ocean. It is a toothed whale with a long, beak-like snout. Bottlenose dolphins are gray to gray-green or gray-brown in color and have a white or pink underbelly.

Bottlenose dolphins are found in almost all the seas worldwide. Adult bottlenose dolphins weigh around 650 kg and are about 2.5 to 3.8 m long. On average, male bottlenose dolphins are larger and heavier than females.

Anatomy

Bottlenose dolphins have a sleek, streamlined body and soft skin. They have a rounded head, and their jaws are positioned in such a way that they appear to have a "smile" on their face. Bottlenose dolphins breathe through a blowhole which is located on their head, and when not in use, remains covered by a muscular flap.

Fins and flukes

Bottlenose dolphins have a sickle-shaped dorsal fin at the center of the back and broad, curved flukes. Bottlenose dolphins propel themselves forward by moving their fluke up and down.

Prey and Predators

Bottlenose dolphins eat a variety of foods—about 8 to 15 kilograms per day—including small fish, squid, octopus, crab, shrimp, eel, and other small sea animals. Bottlenose dolphins have many predators, such as large sharks, including tiger and dusky sharks, and some killer whales. Bottlenose dolphins generally swim in packs to defend themselves against these predators.

Communication

To communicate, bottlenose dolphins whistle, groan, moan, and squeak, and each dolphin has a signature sound that distinguishes it from the others. Bottlenose dolphins also use body language to communicate, splashing and slapping their tail and body and often clapping their jaws together in conflict situations.

Habitat and Range

Bottlenose dolphins inhabit harbors, lagoons, bays, estuaries, and river mouths. There are two known species of bottlenose dolphins: the common bottlenose and the Indo-Pacific bottlenose. The common bottlenose dolphin is found in most warm to tropical oceans, and the Indo-Pacific species is found in the waters around India, Australia, and South-China.

Bottlenose Dolphin Fights

Male bottlenose dolphins often fight with each other to establish supremacy, and these fights are known as "headbutting fights." Bottlenose dolphins are also known to fight to protect themselves and their calves from predators.

Calves

Bottlenose dolphin calves are darker than the adults and have light-colored vertical lines on their sides, which disappear within six months. Calves are between 1.06 and 1.32 meters in length at birth, nurse from their mother for 12 to 18 months, and stay with their mother for anywhere from three to six years.

Bowhead Whale

The bowhead whale is a large baleen whale found in the Arctic seas. Bowheads are blue-black in color with white markings on their belly. Their lower jaw is also white in color and contains a series of irregular, black spots.

Adult bowheads are large, robust whales with white tails. They are generally 15 meters long and more than 60,000 kilograms in weight, with the females being slightly larger than the males.

Habitat and Range

Bowhead whales are the only baleen whales to spend the entirety of their life in icy Arctic seas. Some are found in the Bering Sea, while others, in search of food, move towards the Beaufort Sea in spring and summertime.

Fins, Flippers, and Flukes

Bowhead whales do not have a dorsal fin. They have a deeply-notched fluke, measuring eight meters from tip to tip, and broad, paddle-shaped flippers around 1.8 meters long.

Behavior

Bowhead whales are slow swimmers that either travel alone or in small groups of six individuals; however, larger groups are sometimes seen in feeding areas. Bowhead whales are shy, nonagressive, and social animals who communicate with each other by making a variety of sounds and by slapping their tail and breaching.

Anatomy

Bowhead whales have a large, bow-shaped head with a wide, arch-shaped mouth. In fact, they have the largest mouth and head in the animal kingdom, with their head covering almost two-thirds of their body. Bowhead whales are well adapted to life in the cold seas: they have strong bones in their skull that help them to break through ice, and their massive bodies have a thick layer of blubber which protects them from the cold and also acts as a source of stored energy.

Food Habits

Bowhead whales are filter-feeders and have 300 to 400 baleen plates which help them to sieve their food as they skim through the water with their mouth open. These whales generally feed on plankton and crustaceans like krill, copepods, and amphipods.

Did You Know?

Bowhead whales are known to live for more than 100 years. Scientists have estimated the age of one bowhead whale to be 211 years!

Calves

Female bowhead whales give birth to a calf every three or four years. Calves are born three to five meters long and with a thick layer of blubber, helping them to survive in the freezing ocean waters. Mother bowhead whales nurse their young for about 9 to 12 months.

Common Dolphin

The common dolphin has a complex crisscross color pattern and a tan or yellowish patch on its sides. Common dolphins have a black to brown back with a V-shaped pattern behind the dorsal fin and a white belly. The short-beaked dolphin is more colorful than the long-beaked variety.

Common dolphins display a wide variety of size, shape, and color combinations. There are two species of common dolphins: the long-beaked common dolphin and the short-beaked common dolphin.

Behavior

Common dolphins are highly social animals. They are fast swimmers and great acrobats, often seen lobtailing, breaching, and doing flipper slaps. Common dolphins usually live in pods of up to 100 individuals, but sometimes a pod may even have up to one thousand members. Common dolphins migrate year round, following a circular route from the Mediterranean to the coasts of America and back.

Fins, Flippers, and Flukes

Common dolphins have a triangular or curved dorsal fin which is black to grayish-white in color with a black border. They have thin flukes and pointed flippers.

Habitat and Range

Common dolphins are found throughout the oceans of the world, off the coasts of Africa, southeast Asia, Australia, New Zealand, Japan, and in the Indian Ocean. Long-beaked common dolphins live in coastal waters and short-beaked common dolphins live in offshore waters.

Prey and Predators

Common dolphins feed on a variety of fish including cod, herring, and mackerel and also eat squid and crustaceans such as shrimp and crab. Common dolphins are predated upon by killer whales and are also hunted by humans as a source of food.

Vocalization

Common dolphins are very vocal, often making shrill, shrieking noises that can even be heard above the surface of the water. These dolphins are known to produce around 19 different whistle sounds, which they usually make to communicate with each other and to express stress.

Did You Know?

Common dolphins can dive to depths of 280 meters and stay submerged for up to eight minutes.

Anatomy

Common dolphins have a streamlined body, a long, slender beak, and a single blowhole. Short-beaked common dolphins have a more rounded head than their long-beaked relatives.

Did You Know?

Dusky dolphins can make 40 to 50 acrobatic leaps one after another.

Dusky Dolphin

The dusky dolphin is a small dolphin with a spindle-shaped body. Dusky dolphins are around 1.5 to 2 meters long and about 100 kilograms in weight.

Dusky dolphins are dark or blue-black in color with white or pale gray marks on their sides and a white underbelly. They have a light-colored face with a short, dark beak. These dolphins have a sloping forehead, tall dorsal fins, small flukes, and extended pectoral fins.

Groups

Dusky dolphins are social animals, living and even hunting in groups. They are very fond of leaping and playing and are often seen forming large groups in the afternoon in which to interact and play. By evening, the groups split up into smaller packs of 6 to 15 individuals, and these dolphins then spend the night together.

Sounds

Dusky dolphins make loud, directional sounds, being able to produce a variety of noises including squeaks, squeals, clicks, and whistles.

Calves

Dusky dolphins give birth to a calf every two to three years, usually between the months of June and August. Calves weigh around 5 kilograms and are around 0.6 meters long at birth.

Food Habits

Dusky dolphins hunt in large herds of up to 300 individuals, generally preying on large schools of small sea creatures like anchovy, shrimp, and squid.

Behavior

Dusky dolphins are one of the most acrobatic dolphins in the world, being known for their extraordinary leaps and somersaults, which are each often done as a school. Dusky dolphins are fast swimmers as well. They are curious and friendly animals and are often attracted to boats and people.

Anatomy

Dusky dolphins have a medium-sized body. Their head slopes evenly down from the blowhole to the tip of the snout. They have a bluish-black tail and back and a white underside with a whitish-gray color covering the flanks. The tip of their snout and lower jaw are dark in color, and they have a gray area on each side, extending from the eye down to the flipper.

Habitat and Range

Dusky dolphins live in warm and cool temperate waters and in coastal regions. They are found in the southern hemisphere along the coasts of South America, South Africa, Kerguelen Island, southern Australia, and New Zealand.

Gray Whale

The gray whale is a large baleen whale with a long, streamlined body and a narrow, tapered head. Gray whales have gray, mottled skin and generally appear slate blue or marble white.

Gray whales are around 13 to 14 meters long and weigh anywhere from 27,000 to 36,000 kilograms. They are great divers and swimmers being able to stay underwater for up to 15 minutes and swim at a speed of three to six miles per hour.

Food Habits

Gray whales feed on small animals inlcuding crustaceans such as amphipods and tube worms. Gray whales are filter feeders and have a series of 130 to 180 fringed baleen plates hanging from each side of their upper jaw. They draw water into their mouth and filter the out the food by expelling the water and sediments through the plates.

Anatomy

Gray whales have a robust body with a narrow head And a long, rigid snout. Their upper jaw is arched and slightly overlaps the lower jaw. Gray whales have dimples on the upper jaw with stiff hairs on each depression.

Calves

Female gray whales bear a single calf every two to three years. Calves are dark gray to black in color with white distinctive markings and measure 4.5 to 5 meters long.

Fins, Flippers, and Flukes

Gray whales do not have dorsal fins; instead, they have a low hump on their back with a series of small knobs running down to the flukes. Their flippers are paddle shaped, with pointed tips, and their fluke has a deeply-notched center.

Habitat and Range

Gray whales are generally found in the shallow coastal waters of the North Pacific Ocean. In summer, they migrate along the coast of North America from California to the Arctic or from coastal Korea to Siberia. In winter, they return from the cold seas to the warm waters off the coast of California or Korea.

Behavior

Gray whales are very active and are often found in shallow water, spending most of their time spyhopping, lobtailing, and breaching.

Social Groups

Gray whales generally live in small pods of up to three individuals, although some groups are larger. Gray whales do not form long-lasting bonds.

Humpback Whale

The humpback whale is a large whale, known for arching or humping its back as it dives. Humpback whales are usually black or gray on their back, with a white underbelly.

Humpback whales can grow to be 14 to 16 meters long and weigh up to 50,000 kilograms. Humpbacks are baleen whales and do not have any teeth. Like all baleen whales, female humpbacks are slightly larger than their male counterparts.

Groups

Humpback whales usually live alone but may they travel in loose groups of several individuals, often composed of a mother and her calves.

Calves

Humpback whales give birth to a single calf at a time. Calves are around 3 to 4.5 meters long at birth and can weigh up to 907 kilograms. Calves remain with their mother and survive on her milk for about one year, after which it starts eating solid food.

Anatomy

Humpback whales have a stout and rounded body that narrows towards the tail. They have round, bump-like knobs on top of their head and on their lower jaw, and each knob has at least one stiff hair on it. Humpback whales have two blowholes and a thick layer of blubber which acts as stored energy for the whales to live off of during winter.

Predators

Humpback whales are hunted by killer whales and also by humans, who hunt them for their oil and meat.

Food Habits

Humpback whales are filter feeders, meaning that they filter food through their baleen plates. Humpback whales feed on krill, fish including herring and mackerel, and plankton. Humpbacks can eat up to 2,500 kilograms of food each day.

Songs

Male humpback whales sing attract females, often producing complex songs that las for hours. The songs may be long groans, low moans, roaring sounds, trills, or chirps arranged in a definite repeating pattern. Often, all males in a large group will sing together.

Fins, Flippers, and Flukes

Humpback whales have a small dorsal fin which is irregularly shaped and located on the hump. They also have very long flippers and wide, serrated flukes with pointed tips.

Habitat and Range

Humpback whales are found throughout the world, inhabiting a variety of locations from temperate to tropical waters, including the North Atlantic, North Pacific, and the Southern Ocean. During summer, they migrate to temperate and polar waters for food, and in the winter, they travel to tropical waters to give birth.

Killer Whale

The killer whale or the orca is the largest members of the dolphin family. Orcas are one of the most powerful predators in the world. They grow upto 6 to 7 m long and weigh around 4,000 to 10,000 kilograms.

Killer whales have a striking color pattern: a shiny, black body with a contrasting white chin and underbelly. Killer whales also have a gray patch —which is unique on each animal—just behind the dorsal fin. They also have an oval, white eyepatch behind and above each eye.

Fins and Flippers

Killer whales have a large, triangular dorsal fin which is the tallest in the cetacean world. The fin of males is around 1.8 m tall; females have smaller and slightly curved dorsal fins. Killer whale flippers are broad, rounded, and paddle-shaped and can also Grow up to 1.8 m long.

Calves

Killer whales give birth to a single calf every 3 to 10 years. Calves are generally 1.8 to 2.1 meters long and weigh around 181 kilograms. Calves live on their mother's milk for up to two years and only then start to feed on solid food. Killer whales are very protective of their calves, with all of the members of the pod, including the males, taking care of the calves.

Anatomy

Killer whales are heavy and stocky animals with a sleek and streamlined body. Their head is quite round and tapered and contains a large mouth with 46 to 50 conical teeth. Most individuals are characterized by nicks, scratches, and tears on their dorsal fin.

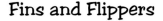

Habitat and Range

Killer whales are found across the oceans of the world; however, they prefer to live in coastal waters and cooler regions. They inhabit the Arctic and Antarctic and sometimes are seen off of the west coast of United States and Canada.

Pods

Killer whales are highly social and cooperative animals, living and traveling in groups of 5 to 30 individuals. A group of killer whales is called a pod, and each pod is lead by females.

Food Habits

Killer whales are skillful hunters and apex predators. They hunt in huge pods, often 40 individuals strong. Killer whales feed on a variety of food such as fish and squid and also hunt large, marine animals like seals, sea lions, penguins, dolphins, porpoises, sea birds, and even other whales.

Did You Know?

Killer whales are fast swimmers, able to travel up to 28 miles per hour when chasing prey.

Minke Whale

The minke whale is a small baleen whale colored gray on its back with a white underside. Minke whales are around eight to nine meters long and can weigh up to 10,000 kilograms. Like all baleen whales, female minke whales are larger than the males.

Minke whales are the most abundant type of whales in the ocean. There are two species: the common, or northern minke and the Antarctic, or southern minke. Common minke whales are slightly larger than their Antarctic minke relatives. The common minke whale includes a subspecies known as dwarf minke whales, which are the smallest of all minkes.

Habitat and Range

Minke whales inhabit almost all of the oceans around the world. They are found generally from the subtropical seas to the polar seas. Some minke whales migrate long distances, while others remain in specific regions throughout the year.

Calves

Female minke whales give birth to a calf every two to three years. Calves are around three meters long and weigh about 450 kilograms. Mother minke whales nurse their calf for up to six months, and after this, the calf may stay alongside its mother until it is about one year old.

Food Habits

Minke whales are filter feeders, and like all other baleen whales, are carnivourous and feed seasonally. The common minke feeds on cod, herring, capelin, pollock, and krill, while the Antarctic minke feeds only on krill.

Fins, Flippers and Flukes

Minke whales have a tall, curved, dorsal fin, short, slender, pointed flippers, and a broad, white band on their dorsal side. The fluke extends into two long tips.

Anatomy

Minke whales are one of the smallest whales in the world. They have a sleek, slender, and streamlined body, a triangular upper jaw, and a pointed, bullet-like head.

Behavior

Minke whales travel alone or in small groups of two to four. They are only found in large groups in areas where there is an abundance of their favorite food, krill.

25

Right Whale

The right whale is a large, heavy baleen whale. It is completely black with only a white patch on its belly. Right whales are one of the rarest whales on Earth.

Right whales are around 12 to 16 meters long and usually weigh about 63,000 kilograms with females being larger than the males. There are three species of right whales: the North Atlantic right whale, the North Pacific right whale, and the Southern right whale.

Anatomy

Right whales have a very unusual appearance. They have a rounded out body, a large head, a a broad back and highly curved—and very hairy—jaws with a long, arching mouth. These whales have large colonies of whale lice that live on their skin, and the colonies appear as white patches on the body of right whales.

Calves

Right whales generally give birth to a single calf at a time, but in very rare cases, twins are born. Newborn calves are around four meters long and stay with their mother for almost a year, during which they feed on her milk.

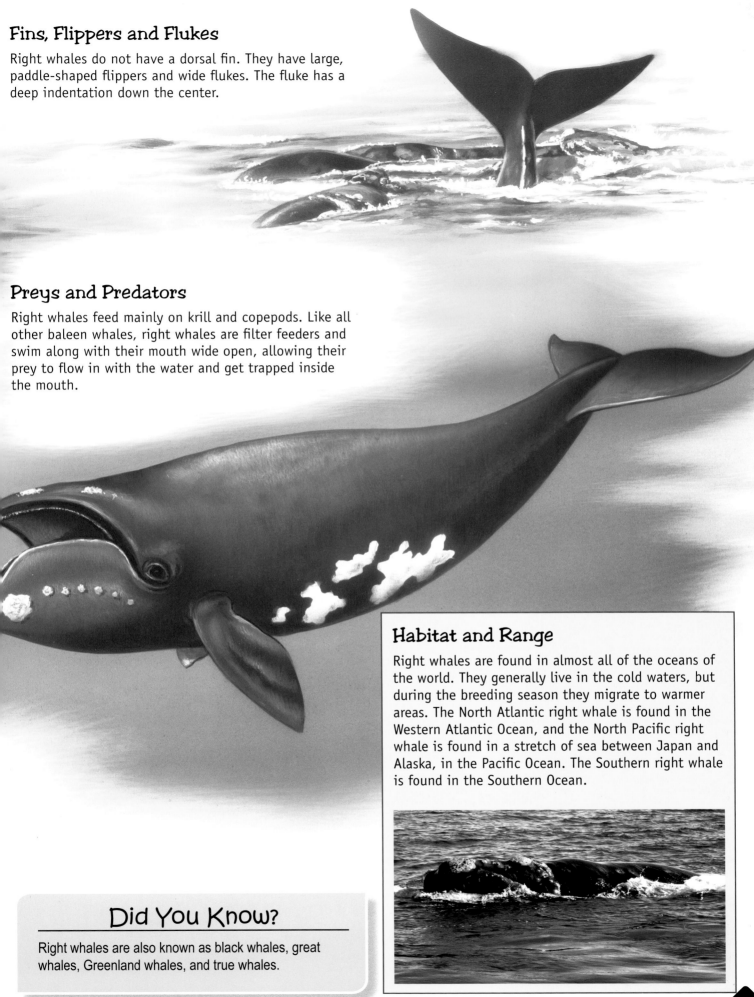

Fins, Flippers and Flukes

Right whales do not have a dorsal fin. They have large, paddle-shaped flippers and wide flukes. The fluke has a deep indentation down the center.

Preys and Predators

Right whales feed mainly on krill and copepods. Like all other baleen whales, right whales are filter feeders and swim along with their mouth wide open, allowing their prey to flow in with the water and get trapped inside the mouth.

Habitat and Range

Right whales are found in almost all of the oceans of the world. They generally live in the cold waters, but during the breeding season they migrate to warmer areas. The North Atlantic right whale is found in the Western Atlantic Ocean, and the North Pacific right whale is found in a stretch of sea between Japan and Alaska, in the Pacific Ocean. The Southern right whale is found in the Southern Ocean.

Did You Know?

Right whales are also known as black whales, great whales, Greenland whales, and true whales.

Sperm Whale

The sperm whale is the largest of the toothed whales. It has an enormous square head and a narrow lower jaw, which make up more than one-third of its length. Sperm whales are dark blue-gray or brownish and have white patches on the belly.

Sperm whales are the largest toothed animals in the world and have a long row of teeth in the lower jaw. Adult male sperm whales measure 15 to 18 meters in length and weigh between 35,000 to 45,000 kilograms. Female sperm whales are much smaller, measuring about 11 meters long and weighing only 13,000 to 14,000 kilograms.

Food Habits

Sperm whales can eat 1,000 kilograms of food each day. They feed primarily on deep water squid, but also eat fish, shrimp, crab, octopus, and even small, bottom-dwelling sharks.

Sperm Whale Groups

Sperm whale females form groups known as family groups or pods. Each pod consists of 10 to 20 female sperm whales and their calves. The calves are taken care of by all of the females in the pod. Sometimes, large males also join the pod for a few hours.

Calves

Sperm whales give birth to only one calf at a time. Calves are about four meters long at birth and weigh around 1,000 kilograms. After birth, newborn calves instinctively swim to the surface within 10 seconds for their first breath, helped by the mother, using her flippers. Within 30 minutes, they begin to swim on their own. The mother nurses the calf for two to three years.

Fins, Flippers and Flukes

Sperm whales typically have a short dorsal fin with knuckles along the spine. They have small flippers that are slightly tapered, and their flukes are broad, triangular, and very thick.

Deepest Divers

Sperm whales are the deepest divers among all the great whales, being able to dive to depths of more than 1,000 meters. Typically, sperm whales dive to depths of around 400 meters and stay underwater for 30 to 45 minutes. During deep dives, they even are known to remain underwater for more than two hours. At this depth, there is little or no light, so sperm whales use echolocation making continuous clicking sounds as they travel through the water, to navigate their way and locate food.

Habitat and Range

Sperm whales are generally found in the higher latitudes and are widely distributed in all of the oceans of the world. Mature males migrate to areas close to Equator, and others migrate to lower latitudes from time to time. The females, newborns, and young whales prefer to stay in the warm tropical and subtropical waters of the Pacific, Atlantic, and Indian Oceans.

Teeth

Sperm whales have 18 to 25 large teeth on each side of the lower jaw. Each cone-shaped tooth can weigh as much as 1 kilogram! The teeth of lower jaw fit into the series of sockets in the upper jaw. Sperm whales only have a few, tiny teeth on the upper jaw.

Brain

The brain of the sperm whale is the largest and heaviest of all animals. The brain of an adult male weighs seven kilograms on average.

Spotted Dolphin

Fins, Flippers and Flukes

Spotted dolphins have a tall, curved dorsal fin and small fl ippers that are pointed. They also have small flukes, which are pointed at the tips and have a small indentation where they join together.

A s its name suggests, the spotted dolphin has spots on its body. It may grow to from 1.7 to 2.4 meters in length and weighs between 90 and 115 kilograms. Female spotted dolphins are much larger than the males.

Spotted dolphins have a tall, curved dorsal fin and small fl ippers that are pointed. They also have small flukes, which are pointed at the tips and have a small indentation where they join together.

Anatomy

Spotted dolphins are small, slender dolphins, with dark skin and a wide, gray stripe. They have a strong body and head with thick, long beaks and white lips. Spotted dolphins have 34 to 48 conical teeth in both the upper and lower jaws.

Pantropical Spotted Dolphin

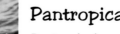

Pantropical spotted dolphins have a long, slender body with a narrow beak. They are found in the tropical and warm waters of the Atlantic, Pacific, and Indian Oceans. Pantropical spotted dolphins feed on a variety of fish including herring and eel and also eat squid, crab larvae, worms, and other invertebrates.

Herds

Spotted dolphins are social animals and live in groups known as herds, which can be made up of hundreds or thousands of dolphins. Pantropical spotted dolphins often live in groups with other species of dolphins, especially spinner dolphins.

Habitat and Range

Spotted dolphins inhabit tropical and warm-temperate waters. They are found in Atlantic Ocean, Pacific Ocean and Indian Ocean.

Calves

Female spotted dolphins give birth to a single calf at a time. Calves measure about 80 to 90 centimeters at birth and grows into young spotted dolphins in about 11 months. The young dolphin then matures into an adult in six to eight years.

Atlantic Spotted Dolphin

Atlantic spotted dolphins are also known as spotted porpoises, Gulf Stream spotted dolphins, and long-snouted dolphins. They are found in the north and south Atlantic. Atlantic spotted dolphins feed on squid, octopus, and variety of fish including herring, flounder, and mackerel.

Food Habits

Spotted dolphins feed on a wide variety of food. They mostly prey on including squid, herring, flounder, mackerel, and anchovies.

Glossary

acrobatic: An art or performance that requires skill and agility.

amphipod: A small crustacean with a vertically thin body and one set of legs for jumping or walking and another set for swimming

anchovy: A small, saltwater fish that comes from the Mediterranean.

baleen: A tough, horny, plate-like structure in the mouth of some whales.

blowhole: The nostril on top of the head of a whale through which it breathes.

blubber: A thick layer of fat below the skin of an animal which keeps it warm in winter.

breaching: To leap clear of the water.

cetacean: An aquatic, hairless mammal of order cetacea, including whales, dolphins; and porpoises.

copepod: A group of small crustaceans found in either salt or fresh water.

dorsal fin: The uppermost fin located between the body and tail.

echolocation: A sensory system in certain animals used to navigate and search for food.

eel: A bony fish with a snake like body found in both salt and fresh water.

flipper: A broad, flat part or limb used for swimming, as in seals and whales.

flounder: A small, flat, bony fish covered with tiny scales.

fluke: The pointed, triangular tail of a whale used for propulsion. The entire fluke is made of two parts which are each called flukes themselves.

indentation: A slight hollow or notch, where the flukes join together.

inlet: A passage or narrow strip of water between an ocean and a bay.

krill: A small crustacean that is the primary food source of baleen whales.

lagoon: A shallow or small lake connected with a larger body of water.

lobtailing: To make a loud splash by forcefully slapping the flukes against the surface of the water.

melon: A large, rounded bulging forehead of a toothed whale containing fatty tissue.

plankton: The microscopic plants and animals that float freely in water.

pollock: A marine food fish having tiny scales; related to cod.

ridge: The long, uppermost section of an animal's back.

rorqual: The baleen whale having a small, pointed dorsal fin and longitudinal grooves on the throat.

rumbling: A deep, heavy, and low-pitched continuous rolling sound.

salmon: A pale red fish with silver scales that lives in both the Atlantic and Pacific Oceans.

sculpin: A small scaleless fish with a long body and wide mouth.

seal: An aquatic carnivore with a torpedo-shaped body and flippers.

sediment: The organic matter that settles at the bottom of a body of water; deposited by wind or water.

smelt: A small, silvery fish found in northern seas or lakes.

somersault: An acrobatic move which is performed by turning the body either forward or backward.

spout: A stream of air or burst of spray from the blowhole of a whale.

spyhopping: To raise the head vertically out of the water.

Repties • Amphibians • Whaes
Whaes and Doiphins • Birds and D
Birds and Ducks • Dianosaurs • Sea Cr
Sea Creatures • Wid Animais
Wid Animais • Insects and Spider • Sharks
Sharks • Mammais • Repties • Amp
Amphibians • Whaes and Doiphins
Birds and Ducks • Dianosa
Wid Animais • Insects and Spider
Sharks • Mammais • Repties
Whaes and Doiphins
Birds and Ducks • Dianosaurs
Dianosaurs • Sea Creatures
Whaes and Doiphins • Birds and Ducks • Dianos
Dianosaurs • Sea Creatures